P9-ECI-754

Sun, Moon, and Planets

Developed at
The Lawrence Hall of Science,
University of California, Berkeley
Published and distributed by
Delta Education,
a member of the School Specialty Family

© 2013 by The Regents of the University of California. All rights reserved. No part of this book may be reproduced or transmitted in any form or by any means, electronic or mechanical, including photocopying or recording, or by any information storage and retrieval system, without permission in writing from the publisher.

1325252
978-1-60902-044-6
Printing 2 — 12/2012
Quad/Graphics, Versailles, KY

Table of Contents

Changing Shadows

Objects you can't see through, such as people, birds, buildings, balls, and flagpoles, have **shadows** on sunny days. That's because opaque objects block sunlight. A shadow is the dark area behind an opaque object. Shadows give information about the position of the **Sun**. If you see your shadow in front of you, then the Sun is behind you.

Did you know that a shadow tells you what time of day it is? Let's see how that works. First, if you are in North America, you need to be facing south. When you are facing south, north is behind you.

It is 12:00 noon. Let's look at the flagpole and observe its shadow. What direction is the shadow pointing?

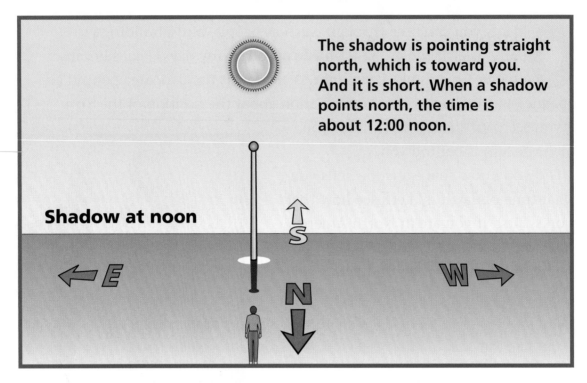

The shadow is pointing straight north, which is toward you. And it is short. When a shadow points north, the time is about 12:00 noon.

Shadow at noon

What does the flagpole's shadow look like at 9:00 in the morning? Did you see your shadow this morning? Do you remember what direction it was pointing? Do you remember how long it was?

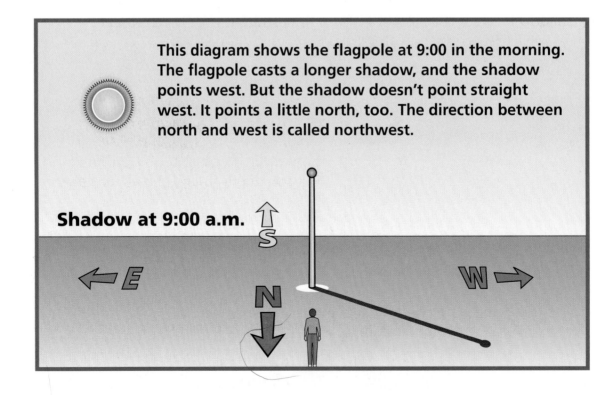

This diagram shows the flagpole at 9:00 in the morning. The flagpole casts a longer shadow, and the shadow points west. But the shadow doesn't point straight west. It points a little north, too. The direction between north and west is called northwest.

Shadow at 9:00 a.m.

What happens to the shadow in the afternoon?

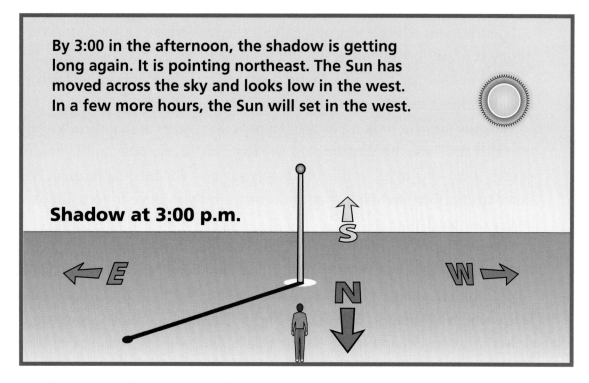

By 3:00 in the afternoon, the shadow is getting long again. It is pointing northeast. The Sun has moved across the sky and looks low in the west. In a few more hours, the Sun will set in the west.

Shadow at 3:00 p.m.

S

E

W

N

Two things happen to a shadow between sunrise and sunset: its length changes and its direction changes. Early in the morning, a shadow is long, and it points west. We observe that the Sun moves across the sky from east to west, and the shadow changes.

At noon, the Sun reaches its highest point in the sky. Now the shadow is as short as it will get. It points straight north. We observe that the Sun keeps moving across the sky. Just before sunset, a shadow is very long, and it points east.

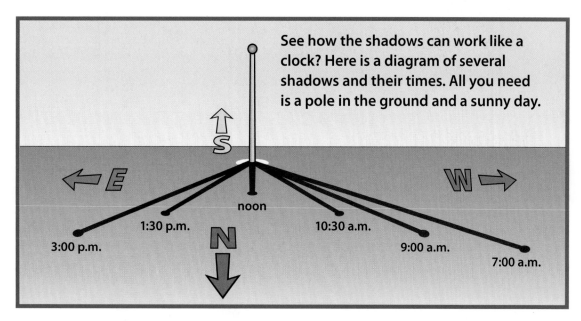

See how the shadows can work like a clock? Here is a diagram of several shadows and their times. All you need is a pole in the ground and a sunny day.

S

E

W

N

noon

1:30 p.m.

10:30 a.m.

3:00 p.m.

9:00 a.m.

7:00 a.m.

The Sun and Seasons

Shadows can tell us even more about the movement of the Sun. We know the Sun moves across the sky from east to west every day. But did you know that the Sun also changes position in the sky from **season** to season? Here's how you can observe it.

Imagine you are looking at the flagpole on your school grounds again. But this time you are standing on the east side of the pole and facing west. North is to your right, and south is to your left. For this experiment, you have to measure the shadow at noon only. But you have to measure it every day for 1 year!

Here are the noon shadows for just five times during the year. Look at the length of the shadow and the position of the Sun in the sky on each date.

On June 21, the first day of summer, the Sun is high in the sky at noon. Three months later, on September 21, the first day of fall, the Sun is lower. And on December 21, the first day of winter, the Sun is at its lowest noon position. After December 21, the Sun begins to climb higher in the sky again. On March 21, the first day of spring, it is as high as it was in September. One year after starting the experiment, on June 21, the Sun is again at its highest noon position.

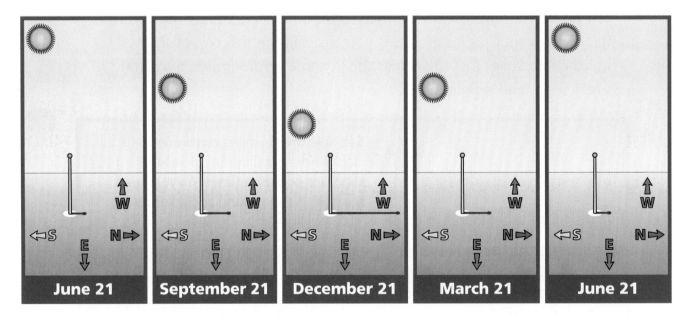

The Sun's change of position in the sky minute by minute during a day is predictable. The Sun's position in the sky season to season during a year is also predictable.

Thinking about Shadows

1. How does the Sun's position in the sky change over 1 day?

2. In what ways do shadows change during the day?

3. What causes shadows to change during the day?

4. Think about a flagpole. How does its shadow change over 1 year?

5. Look at the photo at the top of the page. Can you see the shadow of the person? Can you see the shadows of the four flagpoles? Why or why not?

The Sun rising over a cornfield in Minnesota

Sunrise and Sunset

The Sun has just come up in this picture. It is sunrise. What direction are you looking?

The Sun always rises in the east. If you are in Portland, Maine, the Sun rises in the east. If you are in Portland, Oregon, the Sun rises in the east. If you are in Raleigh, North Carolina, the Sun rises in the east. If you are in Brownsville, Texas, or Broken Bow, Oklahoma, the Sun rises in the east. Wherever you are on **Earth**, the Sun rises in the east.

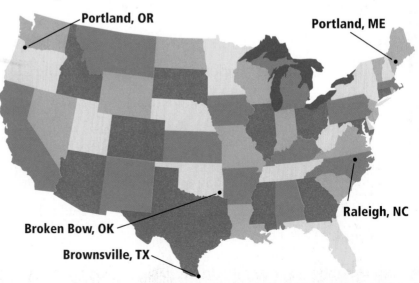

Portland, OR

Portland, ME

Raleigh, NC

Broken Bow, OK

Brownsville, TX

In this picture, the Sun is just about to go down. It is sunset. What direction are you looking now?

That's right, you're looking west. The Sun always sets in the west. If you are in Portland, Maine, the Sun sets in the west. If you are in Portland, Oregon, the Sun sets in the west. If you are in Raleigh, North Carolina, the Sun sets in the west. If you are in Brownsville, Texas, or Broken Bow, Oklahoma, the Sun sets in the west. Wherever you are on Earth, the Sun sets in the west.

Every day the Sun rises in the east and sets in the west. To get from east to west, the Sun appears to slowly travel across the sky. In the early morning, when the Sun first comes up, it is touching the horizon in the east. At noon, the Sun is at its highest position in the sky. At sunset, the Sun is touching the horizon in the west. The Sun's position in the sky changes all day long.

The Sun setting over the city of Boston, Massachusetts

Earth's Rotation

The Sun looks as though it moves across the sky. But it really doesn't. It is Earth that is moving. Here's how it works.

Earth is spinning like a top. It takes 1 day (24 hours) for Earth to **rotate** once. Because Earth is rotating, half of the time we are on the sunny side of Earth. We call the sunny side **day**. The other half of the time we are on the dark side of Earth. We call the dark side **night**.

Imagine it's just before sunrise. You can't see the Sun because you are still on the dark side of Earth. But in 5 minutes, Earth will rotate just enough for you to see the Sun come over the horizon. That moment is sunrise.

Earth turns toward the east, the direction of the orange arrow. That means the first sunlight of the day will be in the east. And, of course, Earth keeps turning. You keep moving with it. In 4 or 5 hours, you have turned so far that the Sun is high over your head. And 5 hours after that, the Sun is low in the western sky. This is because Earth is moving in an eastward direction. It looks as though the Sun is moving across the sky in a westward direction. Finally, it is sunset. The Sun slips below the horizon in the west. It is dark again.

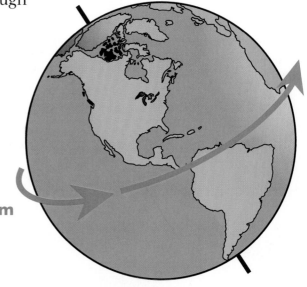

Earth turns toward the east. So the Sun seems to move from east to west across the sky.

The x shows your position just before sunrise.

The x shows your position just after sunrise.

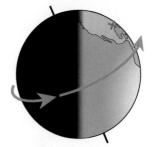

The x shows your position near noon.

There is one thing you can depend on for sure. The Sun will come up tomorrow morning. And you can be sure it will come up in the east. At the end of the day, it will set in the west. You can count on it.

As the day goes along, it looks as though the Sun travels across the sky from east to west. During the morning, it rises higher and higher in the sky. At noon, it is at its highest position in the sky. From noon to sunset, the Sun continues to travel west. And it gets lower and lower in the sky. At sunset, the Sun disappears below the horizon in the west. Another day has passed. And tomorrow will be the same.

Well, almost the same. A careful observer will notice that the Sun's path through the sky is a tiny bit different every day. You can see the difference by studying shadows.

Shadows

A shadow is the dark area behind an opaque object. It is created where an object blocks sunlight. A steel pole, like a flagpole, casts a shadow. The direction of the pole's shadow changes as the Sun's position changes. At noon, the Sun is highest in the sky. Noon is also when the flagpole's shadow is the shortest of the day.

We can watch the noon shadow to see how the Sun's position changes from season to season. The length of that shadow changes a little bit every day. Why does the length of the shadow change? It changes because the position of the Sun at noon changes a little bit every day.

The Sun's position changes all day from sunrise to sunset.

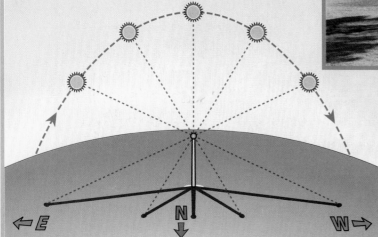

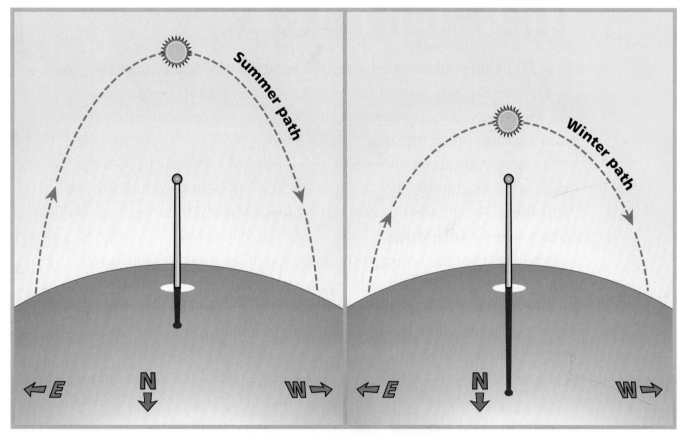

The Sun's path through the sky is higher in summer.

The pattern of change is predictable. In North America, the position of the noon Sun gets higher in the sky from December 21 to June 21. On June 21, the Sun is highest in the sky. That's also the day when the flagpole's shadow is the shortest of the year.

The position of the noon Sun gets lower in the sky each day between June 21 and December 21. On December 21, the noon Sun is lowest in the sky. That's also the day that the flagpole's shadow is the longest of the year.

The Sun's position in the sky changes in two ways. Every day the Sun rises in the east, appears to travel across the sky, and sets in the west. The other way the Sun's position changes is in its daily path. In summer, the Sun's path is high in the sky. In winter, the Sun's path is lower in the sky.

The Night Sky

What do you see when you look up at the sky? During the day, you see the Sun. Sometimes you can see the **Moon**. You might see clouds. If you watch long enough, you will see something fly by, such as a bird or an airplane.

At night, you can see different things in the sky. When the Moon is up, it is the brightest object in the night sky. The Moon might look like a thin sliver. That's called a **crescent Moon**. Or it might be big and round. That's called a **full Moon**.

When you see the Moon in the west, it will set soon. When you see the Moon in the east, it is rising. It is easy to **predict** the time of day or night the Sun will rise and set. It is much harder to predict the time of day or night the Moon will rise and set.

A crescent Moon

The Moon during the day

A full Moon over New York City

14

On a clear night, you can see about 2,000 stars in the sky.

Stars

When it is clear, you can see **stars** in the night sky. Night is the only time you can see stars. Well, almost the only time. There is one star we can see in the daytime. It's the Sun. The Sun shines so brightly that it is impossible to see the other stars. But after the Sun sets, we can see that the sky is full of stars. It looks like there are millions of stars in the sky on a clear night. But actually you can see only about 2,000 stars with your **unaided eyes**.

Venus and Jupiter in the eastern sky just before sunrise

Planets

Some stars are brighter than others. They are the first ones you can see just after the Sun sets. Did you ever make a wish on the first star that appears in the evening sky? "Star light, star bright, first star I see tonight. I wish I may, I wish I might, have the wish I make tonight." That star might not be a star at all. The brightest stars are actually **planets**. That's one way you can tell a planet from a star, by how brightly it appears to shine.

Earth **orbits** the Sun with seven other planets and several **dwarf planets**. You can see five planets in the night sky. Venus is one of the planets you might see. Ancient sky watchers called Venus the evening star. It is seen near the western horizon after sunset. Venus was also called the morning star. It is also seen near the eastern horizon just before sunrise. What caused the confusion?

Two planets orbit closer to the Sun than Earth does. Mercury is closest to the Sun. Then comes Venus. Venus takes only 225 days to go around the Sun. Sometimes Venus is positioned where we can see it from Earth just before sunrise as the morning star. A few months later, Venus has traveled to the other side of the Sun. Now it is positioned for us to see it after sunset as the evening star. That's why ancient sky watchers thought Venus was two different stars.

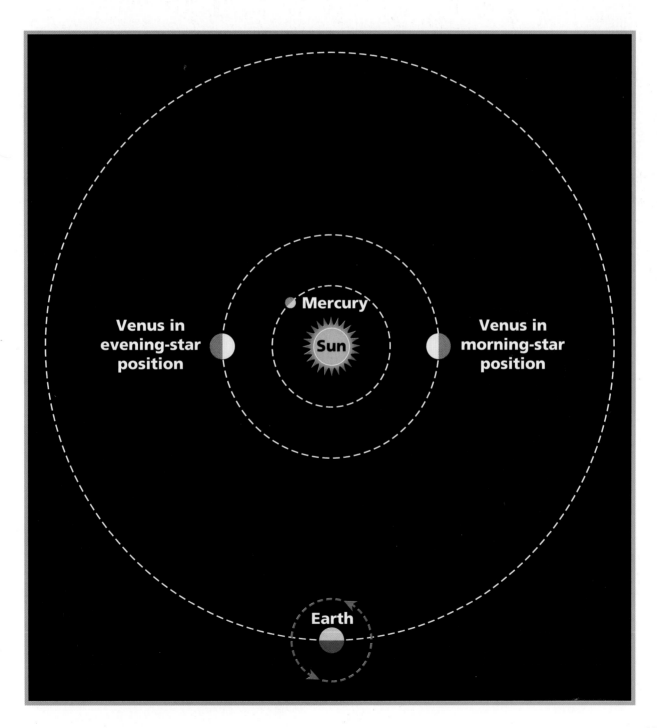

Venus is sometimes visible from Earth.

Four other planets can be seen with unaided eyes. Mercury is visible sometimes. Because it is so close to the Sun, it is often lost in the bright glare of the Sun. Mars is the fourth planet from the Sun. It shines with a slightly red light. Jupiter and Saturn are the farthest of the visible planets. Still, they are pretty bright because they are so big.

It is a special night when you can see all five planets together in the night sky. It doesn't happen very often. It happened in 2004. It won't happen again until 2036!

Thinking about the Night Sky

1. What are some of the objects you can see in the night sky that you can't see in the day sky?

2. Which object is the brightest object in the night sky?

3. Which star is the closest to Earth?

4. Look at the picture of the crescent Moon. What is the other bright object you can see in the night sky?

Comparing the Size of Earth and the Moon

Apollo 11 Space Mission

On July 16, 1969, the world's most powerful booster rocket thundered off launch pad 39A at Cape Canaveral, Florida. Perched on top of the mighty Saturn 5 rocket was a tiny command module and a smaller, spindle-legged, lunar module. The mission was Apollo 11. On board were three American astronauts. Neil Armstrong was the mission commander. Michael Collins was the command module pilot. Edwin "Buzz" Aldrin Jr. was the lunar module pilot.

The goal of the Apollo 11 mission was to land two men on the Moon and return them safely to Earth. The mission was complex. It involved the development of many new technologies, including some of the most advanced engineering ever attempted by humans. The 36-story-tall Saturn 5 three-stage rocket was the largest, most powerful booster rocket ever designed.

The first stage of the rocket lifted the 3,000-ton spacecraft off Earth's surface. After 8 minutes, the first stage was used up and fell away. At that point, the second stage fired up to propel the spacecraft into orbit 189 kilometers (km) above Earth's surface. After orbiting Earth one and a half times, the third stage of the booster rocket fired up and sent the spacecraft on its way toward the Moon.

**Neil Armstrong,
Michael Collins,
and Buzz Aldrin**

The lunar module after it is separated from the command module. The two parts are shown in orbit around the Moon.

The lunar module as it approaches the command module for docking and the return trip to Earth. Earth is seen in earthrise.

As soon as the spacecraft was up to speed, it separated from the third rocket stage and coasted its way to the Moon. Four days later, the spacecraft arrived and moved into orbit around the Moon.

The spacecraft had two separate parts. The first part was the lunar module, the craft that would land on the Moon and later take off from the Moon. The second part was the command module, the craft that would orbit the Moon while waiting for the lunar module to return. The two parts would undock, or separate, during an orbit around the Moon.

When all was ready, Armstrong and Aldrin moved into the lunar module. Mission Control in Houston, Texas, gave the command to the lunar module to start its descent toward the Moon's surface. The two modules separated. *Eagle*, the lunar module, started its long process of slowing down and descending to the Moon's surface. *Columbia*, the command module, stayed in its lunar orbit to await the return of *Eagle* after it completed its mission to the surface.

The preprogrammed descent brought *Eagle* close to the Moon's surface. As *Eagle* approached the landing site, Armstrong and Aldrin could see that they were headed for a pile of boulders. At the last minute, Armstrong took the controls to pilot *Eagle* to a safer landing spot.

Footprints left by the astronauts on the Moon are permanent.

Armstrong took this picture of Aldrin. What can you see in the visor of Aldrin's helmet?

After a few tense seconds, Armstrong guided *Eagle* to a soft, safe landing on the southwestern edge of the Sea of Tranquility. Soon after, Armstrong and Aldrin reported to Mission Control in Texas: "Houston, Tranquility Base here. The *Eagle* has landed!" Dozens of technicians at Mission Control cheered for this amazing event. Humans had arrived safely on the surface of the Moon.

After checking all systems in the lunar module to make sure it was secure and undamaged, Armstrong and Aldrin dressed for a trip outside. The Moon's surface, with no **atmosphere**, is a deadly place for a person without proper protection. The temperature is more than 115 degrees Celsius (°C) in the sunshine and –173°C in the shade. The pressure is 0, and there is no air.

Dressing involved putting on a pressurized space suit that was temperature controlled. The suit provided air and communication. The helmet had a gold-covered lens that could be lowered to protect the astronauts' eyes from dangerous ultraviolet rays from the Sun.

At 10:39 p.m. eastern daylight time, Armstrong squeezed out of the exit hatch onto the ladder leading down to the Moon's surface. As he hopped from the lowest rung onto the Moon's surface, he said, "That's one small step for man, one giant leap for mankind."

The lunar soil onto which Armstrong stepped was like powder. The bulky, stiff suit worked perfectly. Armstrong was comfortable and able to move around easily. Aldrin joined him on the Moon's surface, and together they began their tasks. They set up several experiments on the surface. They put up an American flag, took photos of the terrain, and collected samples of lunar rocks and soil.

After 2 hours and 21 minutes of exploring the Moon's surface, the astronauts gathered their equipment and scientific samples, including 108 kilograms (kg) of Moon rocks, and returned to the lunar module. They repressurized the cabin and settled in for some much needed rest before leaving the Moon's surface.

Aldrin climbs down to step on the Moon's surface.

Armstrong and Aldrin left the American flag on the Moon.

After 7 hours of rest, Mission Control sent the astronauts a wake-up call. Two hours later, they fired the ascent rocket that propelled the lunar module upward. *Eagle* reunited with *Columbia*, which had been orbiting while *Eagle* was on the Moon's surface.

Once the two spacecrafts were reunited, the landing crew transferred to *Columbia*. No longer needed, *Eagle* was left behind in lunar orbit and probably crashed into the Moon in the next few months. Then *Columbia* used its rockets to start its voyage back to Earth. After the long ride home, *Columbia* moved into Earth's orbit. When the time and location were right, rockets fired to push *Columbia* out of orbit and into Earth's atmosphere.

Soon after, huge parachutes opened to slow *Columbia*'s reentry. The historic mission came to a successful end on July 24, when *Columbia* splashed down safely in the Pacific Ocean. They landed only 24 km from the recovery ship waiting for their return.

Six more Apollo missions followed this adventure. The last mission, Apollo 17, was in December 1972. A total of 12 people have walked on the Moon. The Moon is the only **extraterrestrial** object that humans have visited.

Columbia **safely landed in the Pacific Ocean.**

Changing Moon

Earth has one large **satellite**. It is called the Moon. The Moon completes one orbit around Earth every 28 days. One complete orbit is also called a **cycle**.

The Moon is the second-brightest object in the sky. It shines so brightly that you can see it even during the day. But did you know that the Moon doesn't make its own light? The light you see coming from the Moon is **reflected** sunlight. Sunlight reflected from the Moon is what we call moonlight.

The Moon is a sphere. When light shines on a sphere, the sphere is half lit and half dark. Wherever you position the sphere, if the light source is shining on it, one half will always be lit and the other half dark.

The same is true for the Moon. It is always half lit and half dark. The half that is lit is the side toward the Sun. The half that is dark is the side away from the Sun.

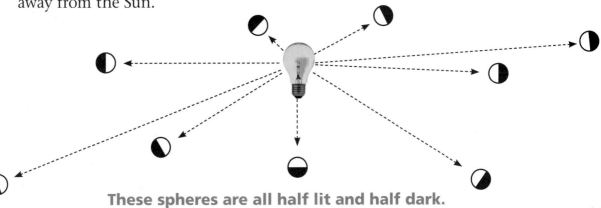

These spheres are all half lit and half dark.

25

The Moon's Position

From Earth, the Moon never looks the same 2 days in a row. Its appearance changes all the time. Sometimes it looks like a thin sliver, and sometimes it looks completely round. Why does the Moon's appearance change?

The Sun is in the center of the **solar system**. The planets orbit the Sun. The Moon orbits Earth.

It takes 4 weeks for the Moon to orbit Earth. Where is the Moon during those 4 weeks? Let's take a look from out in space.

We'll start the observations when the Moon is at position 1 between Earth and the Sun.

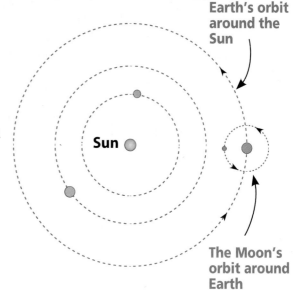

Earth's orbit around the Sun

The Moon's orbit around Earth

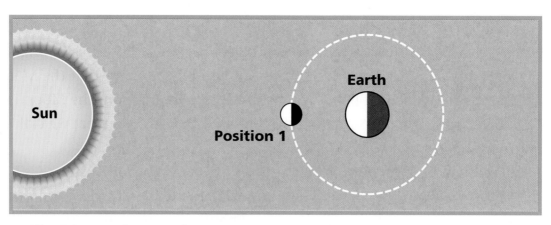

The Moon orbits Earth in a counterclockwise direction. After 1 week, the Moon has moved to position 2.

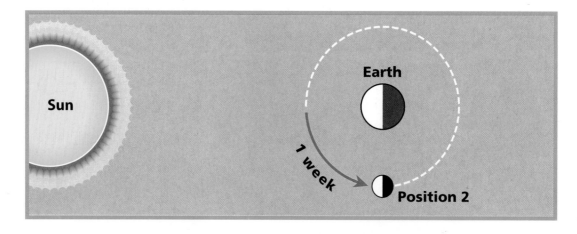

After 2 weeks, the Moon has moved to position 3, on the other side of Earth. The Moon has traveled halfway around Earth.

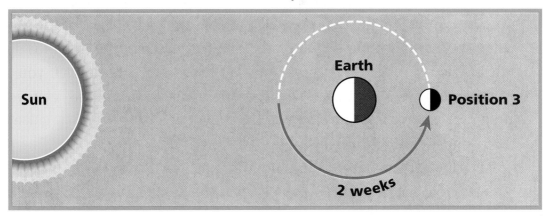

After 3 weeks, the Moon has moved to position 4. It is now three-quarters of the way around Earth.

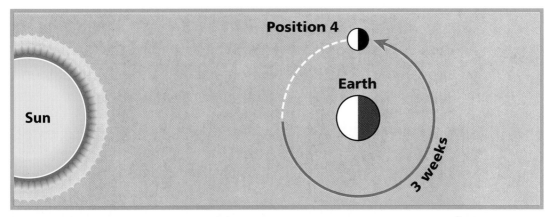

In another week (a total of 4 weeks), the Moon has returned to position 1. It has completed one **lunar cycle**.

Look at the Moon in each of the illustrations. You will see that the lit side is always toward the Sun. In each position during the lunar cycle, the Moon's bright side is always toward the Sun.

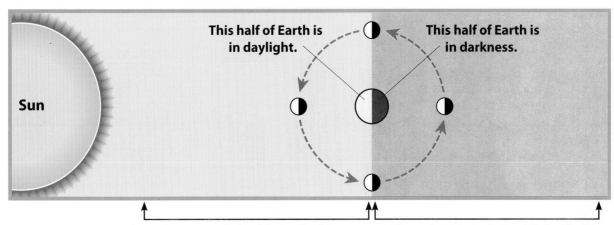

The Moon's Appearance

The shape of the Moon doesn't change. It is always a sphere. The amount of the Moon that is brightly lit doesn't change. Half of the Moon is always lit by the Sun. What changes is how much of the lit half is visible from Earth. You might see just a tiny bit of the lit half. Or you might see the entire lit half. The lit portion you see from Earth changes in a predictable way. The different shapes you see have been named, and each one is called a **phase**.

Let's look at the phase of the Moon in position 1. The Moon is between the Sun and Earth. When you look up at the Moon from Earth, what do you see? Nothing. All of the lit half of the Moon is on the other side. This is the **new Moon**. The new Moon has no light visible from Earth. The new Moon is shown as a black circle.

Let's move forward 2 weeks. The Moon has continued in its orbit and is in position 3. What do you see when you look up at the Moon? The whole lit side of the Moon is visible from Earth. This is the full Moon. The full Moon is shown in the illustration as a white circle.

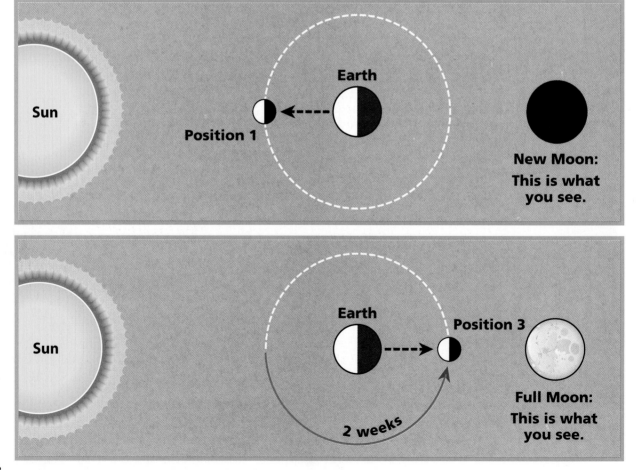

Now let's look at positions 2 and 4. At both positions, you see half of the lit part of the Moon and half of the dark part of the Moon. At position 2, when you look up at the Moon, the lit part is on the right side. At position 4, the lit part of the Moon is on the left side. Position 2 is the **first-quarter Moon**. Position 4 is the **third-quarter Moon**.

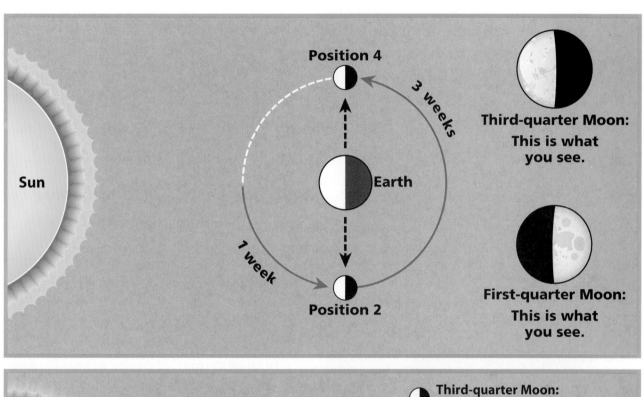

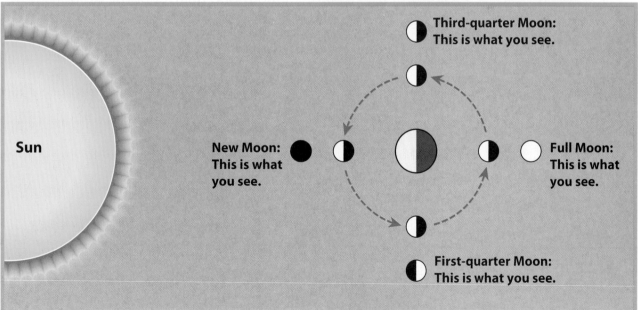

This is the Moon's orbit during one lunar cycle.

Day 0: new Moon

Day 3: waxing crescent Moon

Day 5: waxing crescent Moon

Day 6: waxing crescent Moon

Lunar Cycle

The new Moon is invisible for two reasons. First, no light is coming to your eyes from the Moon. The lit side is facing away from Earth. Second, to look for the new Moon, you would have to look toward the Sun. The glare is too bright to see the Moon. (And remember, you should never look directly at the Sun.)

Three days later, the Moon has moved in its orbit, and it is visible. The first sighting of the Moon after a new Moon is a tiny sliver of visible light. The curved shape is called the crescent Moon.

On day 5, the Moon looks larger. About one-quarter of the Moon is now bright. Each day the visible bright part of the Moon is a little larger. We say the Moon is **waxing** when it appears to be growing.

By day 6, almost half of the Moon appears brightly lit. This is the last day of the waxing crescent Moon. Tomorrow the Moon will appear as the first-quarter Moon.

The first-quarter Moon is the phase seen on day 7. The Moon has completed the first quarter of its lunar cycle. Observers on Earth see half the sunlit side of the Moon and half the dark side of the Moon. The brightly lit side is on the right side.

On day 9, you can see more than half the sunlit side of the Moon. The Moon appears to be oval shaped. A Moon phase that is larger than a quarter but not yet full is called a **gibbous Moon**. Because the Moon is still getting bigger, it is a waxing gibbous Moon.

On day 11, the Moon is almost round. It is still a waxing gibbous Moon. Observers on Earth can see most of the sunlit half of the Moon. They can see only a small sliver of the dark side of the Moon. Can you see the dark crescent?

On day 14, you can see the whole sunlit side of the Moon. This is the full-Moon phase. A full Moon always rises at the same time the Sun sets.

Day 7: first-quarter Moon

Day 9: waxing gibbous Moon

Day 11: waxing gibbous Moon

Day 14: full Moon

Day 18: waning gibbous Moon

Each day after the full Moon, the bright part of the Moon gets smaller. Getting smaller is called **waning**. On day 18, the Moon looks oval again. Because it is still between full-Moon phase and quarter phase, it is still a gibbous Moon, a waning gibbous Moon.

Day 21: third-quarter Moon

On day 21, the Moon has completed three-quarters of its orbit around Earth. The Moon appears as the third quarter, again half bright and half dark. But notice that the bright side of the third-quarter phase is on the left. Compare the appearance of the third-quarter Moon and the first-quarter Moon.

Day 24: waning crescent Moon

As the Moon starts the last 7 days of its orbit, it returns to crescent phase. But because it is getting smaller each day, it is the waning crescent phase. By day 24, an observer on Earth sees just a small part of the sunlit side of the Moon. A lot of the dark side is visible again.

Day 28: new Moon

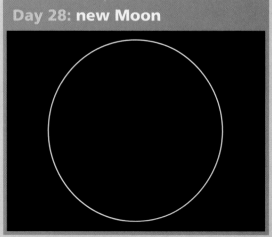

On about day 28, the Moon has completed one lunar cycle. It is back at its starting point. It is at the new-Moon phase again. The night sky is moonless. The day sky has no Moon. For a couple of days, observers on Earth can't see the Moon.

Then, in the evening sky, just after sunset, the Moon reappears. It is a thin, silver-colored crescent. And if you are in the right place at the right time, you could see something special. It is a bright crescent on the edge of a dim full Moon. It is called the old Moon in the new Moon's arms.

How can you see a bright crescent Moon and a pale full Moon at the same time? When the Moon appears as a thin crescent, it is almost between Earth and the Sun. A lot of light reflects from Earth onto the Moon. The whole Moon is dimly lit by earthshine.

The old Moon in the new Moon's arms

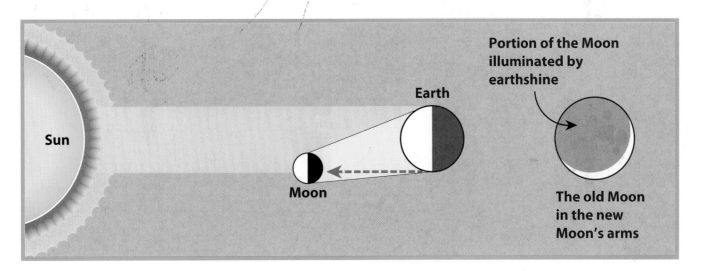

Sun

Earth

Moon

Portion of the Moon illuminated by earthshine

The old Moon in the new Moon's arms

Thinking about the Phases of the Moon

1. How long does it take the Moon to complete one lunar cycle?

2. What is a new Moon, and what causes it?

3. What is the difference between a waxing Moon and a waning Moon?

4. What is the difference between a crescent Moon and a gibbous Moon?

5. Describe the Moon's appearance 1 week, 2 weeks, 3 weeks, and 4 weeks after the new Moon.

Lunar Cycle Diagram

The position of the Moon in its lunar cycle determines the Moon's phase.

Illustrations in this circle show the Moon's location during its 4-week orbit around Earth. Note that half of the Moon is bright at all times because sunlight is shining on the side of the Moon facing the Sun.

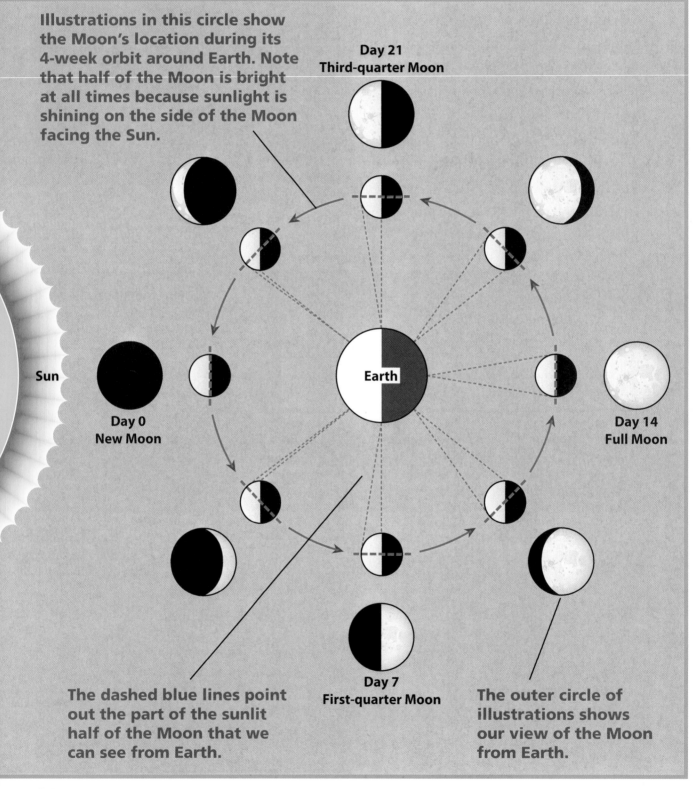

Day 21
Third-quarter Moon

Sun

Earth

Day 0
New Moon

Day 14
Full Moon

Day 7
First-quarter Moon

The dashed blue lines point out the part of the sunlit half of the Moon that we can see from Earth.

The outer circle of illustrations shows our view of the Moon from Earth.

A total
solar
eclipse

A total
lunar
eclipse

Eclipses

Occasionally, people on Earth are able to observe a lovely orange-colored eclipse of the Moon (a **lunar eclipse**). Less frequently, they can observe a black-centered eclipse of the Sun (a **solar eclipse**). What causes these interesting events? When can you see a lunar eclipse? When can you see a solar eclipse?

What Is a Solar Eclipse?

A solar eclipse occurs when the Moon passes exactly between Earth and the Sun. The Moon completely hides the disk of the Sun when this happens. This diagram shows the alignment of Earth, the Moon, and the Sun during a solar eclipse.

A total solar eclipse

You can see the solar eclipse only if you are where the Moon's shadow falls on Earth's surface. A solar eclipse lasts for about 7 minutes.

Earth

Moon

Sun

You can see a solar eclipse only on a very small area of Earth's surface. A total eclipse of the Sun is visible for a bit more than 7 minutes, as long as it takes for the disk of the Moon to pass across the disk of the Sun.

The Moon travels around Earth once every month. Why doesn't a solar eclipse occur every month? The Moon's orbit around Earth is not in the same plane as the orbit of Earth going around the Sun. The Moon's orbit is tilted a little bit. Most months, Earth, the Moon, and the Sun are not in a straight line.

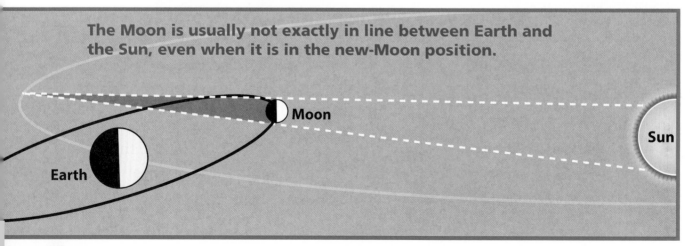

The Moon is usually not exactly in line between Earth and the Sun, even when it is in the new-Moon position.

Moon

Sun

Earth

What Is a Lunar Eclipse?

A lunar eclipse occurs when Earth passes exactly between the Moon and the Sun. Earth's shadow completely covers the disk of the Moon when this happens. This diagram shows the alignment of Earth, the Moon, and the Sun during a lunar eclipse.

A lunar eclipse

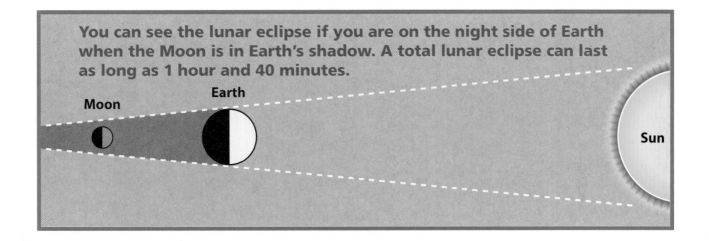

You can see the lunar eclipse if you are on the night side of Earth when the Moon is in Earth's shadow. A total lunar eclipse can last as long as 1 hour and 40 minutes.

Moon   Earth   Sun

You can see a lunar eclipse from anywhere on Earth where it is night. A total lunar eclipse lasts almost 2 hours and its beautiful red color is safe to view without eye protection.

Why don't we see a lunar eclipse every month? Again, it's because of the tilt of the Moon's orbit around Earth. The Moon's orbit around Earth is not in the same plane as the orbit of Earth going around the Sun. In most months, Earth's shadow does not fall on the Moon.

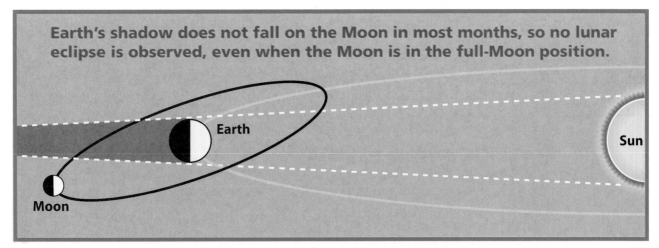

Earth's shadow does not fall on the Moon in most months, so no lunar eclipse is observed, even when the Moon is in the full-Moon position.

Earth   Sun   Moon

Look at the sequence of photos showing a total lunar eclipse. The Moon moves across Earth's shadow. The diagram below the photos explains what's happening. Note the reddish-brown color of the last photo.

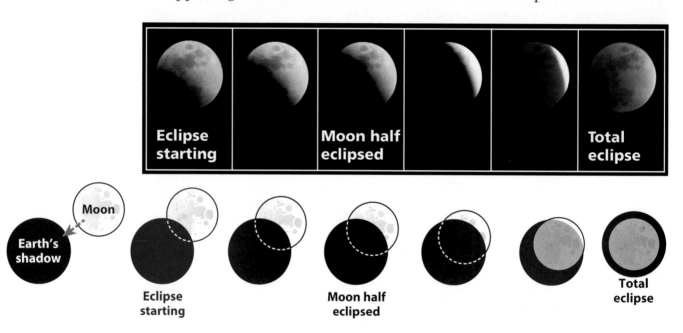

Eclipse starting

Moon half eclipsed

Total eclipse

Moon

Earth's shadow

Eclipse starting

Moon half eclipsed

Total eclipse

Why is the totally eclipsed Moon reddish-brown? Why is it visible at all? When light passes through Earth's atmosphere, it is bent and scattered by the air. As a result, some reddish light falls on the Moon's surface. This makes the Moon appear reddish-brown. If this bending and scattering did not occur, a totally eclipsed Moon would be invisible because no light would hit the Moon. If no light hit the Moon, no light would be reflected back into our eyes on Earth.

Thinking about Eclipses

1. During what phase of the Moon can you observe a lunar eclipse?

2. During what phase of the Moon can you observe a solar eclipse?

Sizes and distances of solar-system objects are not drawn to scale.

Exploring the Solar System

Imagine you are coming to the solar system as a stranger on a tour. There is a tour guide to provide information. You have a window to look out. The tour is about to start. What will you see?

The first view of the solar system is from space. From here, you can see the whole solar system. The most surprising thing is that the solar system is mostly empty. The **matter** is concentrated in tiny dots. And the dots are far apart. Most of the dots are planets.

There is a star in the center of the solar system. Four small planets orbit pretty close to the star. These are the rocky **terrestrial planets**.

Next, there is a region of small bits of matter orbiting the star. This is the **asteroid** belt.

Out farther, four large planets orbit the star. These are the **gas giant planets** made of gas.

Beyond the gas giant planets is a huge region of icy chunks of matter called the **Kuiper Belt**. Some of the chunks are big enough to be planets. A dwarf planet, Pluto, is one of the Kuiper Belt objects. Others have orbits that send them flying through the rest of the solar system.

The Sun

The Sun is a fairly average star. It is much like millions of other stars in the **Milky Way**. The Sun formed about 5 billion years ago. A cloud of gas began to spin. As it spun, it formed a sphere. The sphere got smaller and smaller. As it got smaller, it got hotter. Eventually, the sphere got so hot that it started to radiate light and heat. A star was born.

The Sun is made mostly of hydrogen (72 percent) and helium (26 percent). It is huge. The **diameter** is about 1,384,000 kilometers (km). The diameter is the distance from one side of the Sun to the other through the center. That's about 109 times the diameter of Earth.

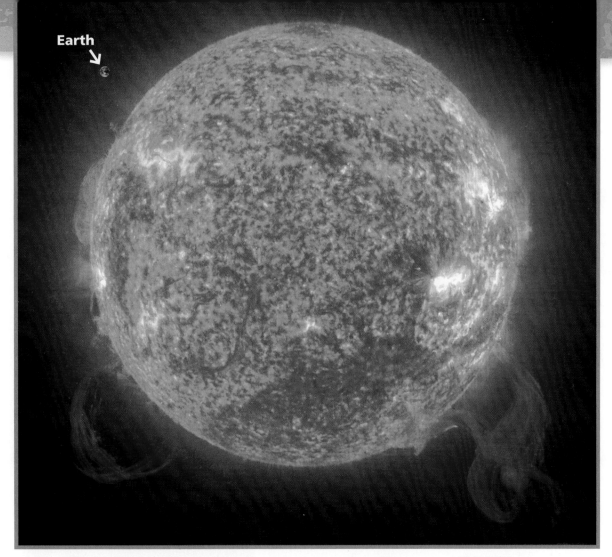

Earth

The Sun's diameter is about 109 times the diameter of Earth.

The Sun is incredibly hot. Scientists have figured out that the temperature at the center of the Sun is 15,000,000 degrees Celsius (°C). The temperature of the Sun's surface is lower, about 5,500°C. Hydrogen is constantly changing into helium in **thermonuclear reactions**. These reactions create heat and light. About 3.6 tons of the Sun's **mass** is being changed into heat and light every second. This energy radiates out from the Sun in all directions. A small amount of it falls on Earth.

Another name for the Sun is Sol. That's why the whole system of planets is called the solar system. The solar system is named for the ruling star. The Sun rules because of its size. It has 99.8 percent of the total mass of the solar system. All the other solar-system objects travel around the Sun in predictable, almost-circular orbits. The most obvious objects orbiting the Sun are the planets.

Terrestrial Planets

The terrestrial planets are the four planets closest to the Sun. They are small and rocky.

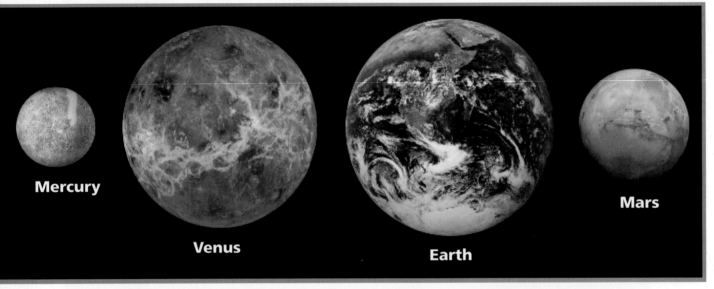

Relative sizes of the terrestrial planets

Mercury

Mercury is the planet closest to the Sun. Mercury is smaller than Earth and has no satellite (moon). By human standards, it is an uninviting place. Mercury is very hot on the side facing the Sun and very cold on the dark side. It has no atmosphere or water.

Mercury is covered with **craters**. The craters are the result of thousands of collisions with objects flying through space. The surface of Mercury looks a lot like Earth's Moon.

Mercury is the planet closest to the Sun.

Venus

Venus is the second planet from the Sun. Venus is about the same size as Earth and has no satellites. The surface of Venus is very hot all the time. It is hot enough to melt lead, making it one of the hottest places in the solar system.

There is no liquid water on Venus. But Venus does have an atmosphere of carbon dioxide. The dense, cloudy atmosphere makes it impossible to see the planet's surface. Modern radar, however, allows scientists to take pictures through the clouds. We now know that the surface of Venus is dry, cracked, and covered with volcanoes.

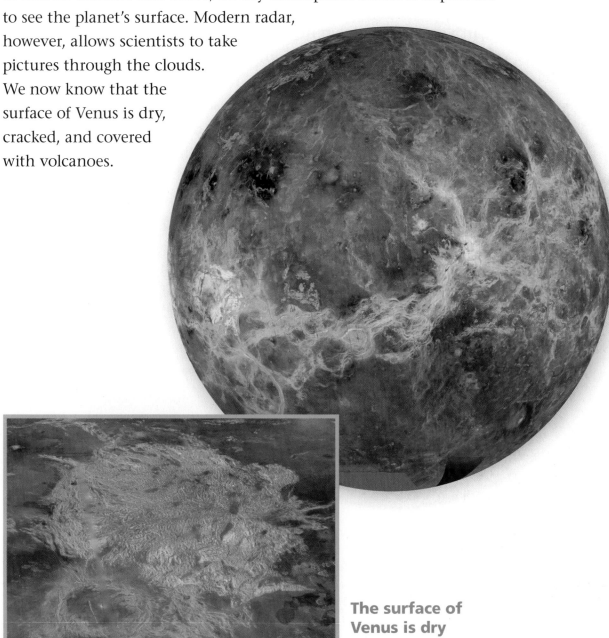

The surface of Venus is dry and covered with volcanoes.

Earth

Earth is the third planet from the Sun. Earth has a moderate, or mild, temperature all the time. It has an atmosphere of nitrogen and oxygen, and it has liquid water. As far as we know, Earth is the only place in the universe that has life. Earth also has one large satellite called the Moon. The Moon orbits Earth once a month. The Moon is responsible for the tides in Earth's ocean. The Moon is the only extraterrestrial place humans have visited.

Earth is 150 million km from the Sun. This is a huge distance. It's hard to imagine that distance, but think about this. Sit in one end zone of a football field and curl up into a ball. You are the Sun. A friend goes to the other end zone and holds up the eraser from a pencil. That's Earth. Get the idea? Earth is tiny, and it is a long distance from the Sun. Still, the light and heat that reach Earth provide the right amount of energy for life as we know it.

The Moon orbits Earth once a month.

Water frost on the surface of Mars

A robotic lander exploring Mars

Mars

Mars is the fourth planet from the Sun. It has two small satellites, Phobos and Deimos. Mars is a little like Earth, except it is smaller, colder, and drier. There are some places on Mars that are like Death Valley in California. Other places on Mars are more like Antarctica, and others are like the volcanoes of Hawaii.

Mars is sometimes called the red planet because of its red soil. The soil contains iron oxide, or rust. The iron oxide in the soil tells scientists that Mars probably had liquid water at one time. But liquid water has not been on Mars for 3.5 billion years. It has frozen water in polar ice caps that grow and shrink with its seasons.

Mars is likely the next place humans will visit. But exploring Mars will not be easy. Humans can't breathe the thin atmosphere of carbon dioxide. And explorers will need to wear life-support space suits for protection against the cold.

Several robotic landers, including *Viking, Spirit, Opportunity, Sojourner,* and *Curiosity* have observed Mars and sent back information about the surface and presence of water. Evidence suggests that there is a lot of frozen water just under the surface.

Asteroids

Beyond the orbit of Mars are millions of chunks of rock and iron called asteroids. They all orbit the Sun in a region called the asteroid belt. The asteroid belt surrounds the terrestrial planets. The planets farther out are quite different from the terrestrial planets.

Some asteroids even have moons. When the spacecraft *Galileo* flew past asteroid Ida in 1993, scientists were surprised to discover it had a moon. They named it Dactyl. The largest object in the asteroid belt is Ceres, a dwarf planet. It is about 960 km around.

Asteroid Ida with moon Dactyl

Gas Giant Planets

The four planets farthest from the Sun are the gas giant planets. They do not have rocky surfaces like the terrestrial planets. So there is no place to land or walk around on them. They are much bigger than the terrestrial planets. What we have learned about the gas giant planets has come from probes on rockets sent out to fly by and orbit around the giants. Even though they are made of gases, each gas giant planet is different.

Neptune

Uranus

Saturn

Jupiter

Relative sizes of the gas giant planets

Jupiter and its four largest moons

Jupiter

Jupiter is the fifth planet from the Sun. It is the largest planet in the solar system. It is 11 times larger in diameter than Earth. Scientists have found 63 moons orbiting Jupiter. The four largest moons are Ganymede, Callisto, Io, and Europa.

Jupiter's atmosphere is cold and poisonous to life. It is mostly hydrogen and helium. Jupiter's stripes and swirls are cold, windy clouds of ammonia and water. Its Great Red Spot is a giant storm as wide as three Earths. This storm has been raging for hundreds of years. On Jupiter, the atmospheric pressure is so strong that it squishes gas into liquid. Jupiter's atmosphere could crush a metal spaceship like a paper cup.

An artist's drawing of Jupiter, its moon Io, and the *Galileo* spacecraft

Saturn

Saturn is the sixth planet from the Sun. It is the second largest planet and is very cold. At least 60 satellites orbit Saturn. Most of the planet is made of hydrogen, helium, and methane. It doesn't have a solid surface.

It has clouds and storms like Jupiter, but they are harder to see because they move so fast. Winds in Saturn's upper atmosphere reach 1,825 km per hour.

The most dramatic feature of Saturn is its ring system. The largest ring reaches out 200,000 km from Saturn's surface. The rings are made of billions of small chunks of ice and rock. All the gas giant planets have rings, but they are not as spectacular as Saturn's.

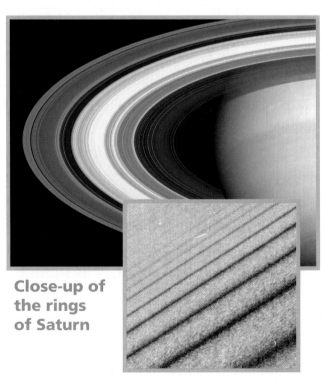

Close-up of the rings of Saturn

Uranus

Uranus is the seventh planet from the Sun. Uranus has 27 moons and 11 rings. Uranus is very cold and windy, and would be poisonous to humans. It is smaller and colder than Saturn.

Uranus has clouds that are extremely cold at the top. Below the cloud tops, there is a layer of extremely hot water, ammonia, and methane. Near its core, Uranus heats up to 4,982°C. Uranus looks blue because of the methane gas in its atmosphere.

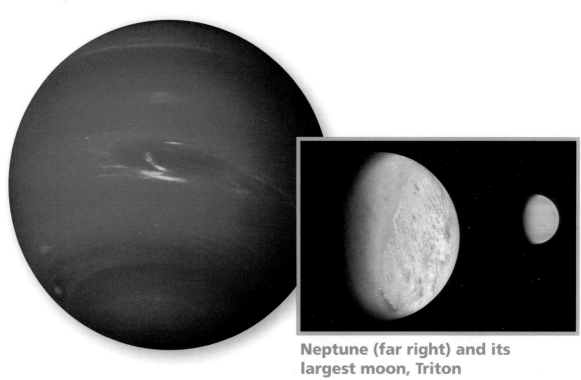

Neptune (far right) and its largest moon, Triton

Neptune

Neptune is the eighth planet from the Sun. Neptune has 13 moons and 4 thin rings. It is the smallest of the gas giant planets, but is still much larger than the terrestrial planets.

Neptune is made mostly of hydrogen and helium with some methane. It may be the windiest planet in the solar system. Winds rip through the clouds at more than 2,000 km per hour. Scientists think there might be an ocean of super-hot water under Neptune's cold clouds. It does not boil away because of the atmospheric pressure.

Pluto and Charon, one of its moons

Kuiper Belt

Out beyond the gas giant planets is a disk-shaped zone of icy objects called the Kuiper Belt. Some of the objects are fairly large.

Pluto

Pluto is a large Kuiper Belt object. Some scientists considered Pluto a planet because it is massive enough to form a sphere. Others did not consider Pluto a planet. To them, Pluto is one of the large pieces of debris in the Kuiper Belt. Scientists have agreed to call Pluto a dwarf planet.

Pluto has a thin atmosphere. When Pluto is farthest from the Sun, the atmosphere gets so cold that it freezes and falls to the surface. Even though Pluto is smaller than Earth's Moon, it has its own moons. Charon is the largest (about half the size of Pluto). Nix and Hydra are much smaller, and in 2011, an even smaller moon named S/2011 P1 was discovered. And there may be more!

Eris

In July 2005, **astronomers** at the California Institute of Technology announced the discovery of a new planet-like object. It is called Eris. Like Pluto, Eris is a Kuiper Belt object and a dwarf planet. But Eris is more than twice as far away from the Sun as Pluto is. This picture is an artist's idea of what the Sun would look like from a position close to Eris.

The Sun would look like a bright star from Eris.

Comets

Comets are big chunks of ice, rock, and gas. Sometimes comets are compared to dirty snowballs. Scientists think comets might have valuable information about the origins of the solar system.

Comets orbit the Sun in long, oval paths. Most of them travel way beyond the orbit of Pluto. A comet's trip around the Sun can take hundreds or even millions of years, depending on its orbit. A comet's tail shows up as it nears the Sun and begins to warm. The gases and dust that form the comet's tail always point away from the Sun.

A comet's tail always points away from the Sun.

Comet orbits can cross planet orbits. In July 1994, a large comet, named Comet Shoemaker-Levy 9, was on a collision course with Jupiter. As it got close to Jupiter, the comet broke into 21 pieces. The pieces slammed into Jupiter for a week. Each impact created a crater larger than Earth.

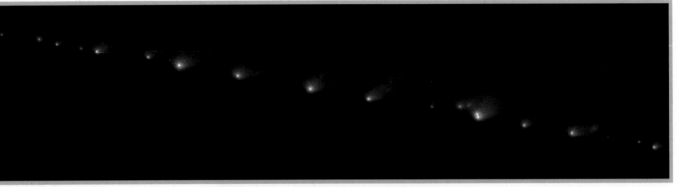

Comet Shoemaker-Levy 9 broke into 21 pieces as it got close to Jupiter.

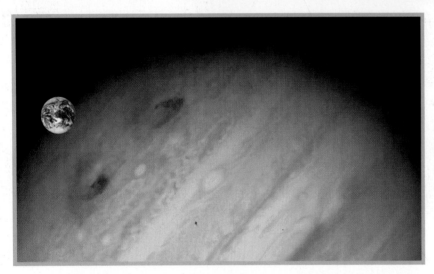

Two of the comet's craters on Jupiter. The picture of Earth gives an idea of how big the craters are.

Thinking about the Solar System

1. What is the Sun, and what is it made of?

2. What is the solar system?

3. Which planets are terrestrial planets? Which planets are gas giant planets?

4. What is the Kuiper Belt, and what is found there?

5. Which planet has the most moons orbiting it?

6. How are asteroids and comets alike and different?

Planets of the Solar System

Mercury

Venus

Earth

Pluto (dwarf planet)

Mars

Sun

Relative sizes of planets

Jupiter

Saturn

Neptune

Uranus

53

Why Doesn't Earth Fly Off into Space?

Earth travels around the Sun in a predictable, almost-circular path once every year. That's a distance of about 942 million kilometers (km) each year. That's an incredible 2.6 million km each day! Earth travels at a speed over 100,000 km per hour. That's fast.

One important thing to know about objects in motion is that they travel only in straight lines. Objects don't change direction or follow curved paths unless a force pushes or pulls them in a new direction. If nothing pushed or pulled on Earth, it would fly off into space in a straight line.

But Earth doesn't fly off into space in a straight line. Earth travels in an almost-circular path around the Sun. In order to travel a circular path, Earth has to change direction all the time. Something has to push or pull Earth to change its direction. What is pushing or pulling Earth? The answer is **gravity**.

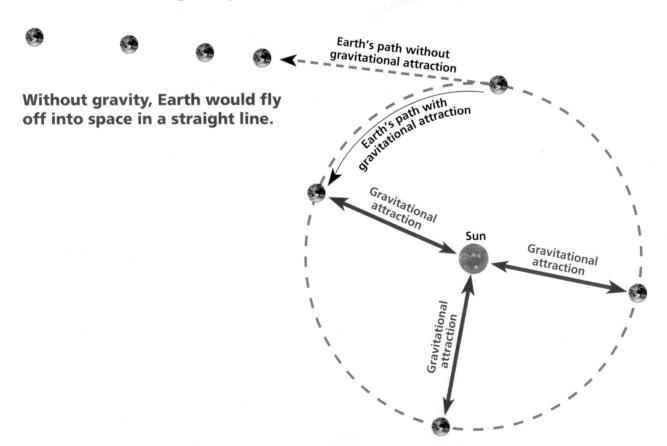

Earth's path without gravitational attraction

Earth's path with gravitational attraction

Without gravity, Earth would fly off into space in a straight line.

Gravitational attraction

Sun

Gravitational attraction

Gravitational attraction

Gravity is the force of attraction between objects. The Sun is an object. Earth is an object. The force of attraction between the Sun and Earth pulls hard enough to change Earth's direction of travel.

Remember the string-and-ball demonstration? The hand pulled on the string. The string pulled on the ball. The ball traveled in a circular orbit. Gravity is like the string. The force of **gravitational attraction** between the Sun and Earth pulls on Earth, changing its direction of travel. The pull of gravity doesn't change Earth's speed, just its direction. That's why Earth travels in an almost-circular orbit around the Sun.

The Sun's gravity keeps all the planets in their orbits. Otherwise, each planet would fly off in a straight line right out of the solar system.

Earth travels around the Sun.

Thinking about Orbits

1. Why do planets stay in orbit around the Sun?

2. How is a ball on a string like a planet in its orbit?

3. What keeps the Moon in its orbit around Earth?

How Did Earth's Moon Form?

Counting out from the Sun, Earth is the first planet with a satellite, or moon. Mercury and Venus, closer to the Sun, don't have moons. Mars, the fourth planet from the Sun, has two moons. Earth probably didn't have a moon at first. Earth got its Moon early in Earth's history as a result of a gigantic collision. The event might have happened about 4.5 billion years ago. This is how it might have happened.

Earth formed from gas and dust in the solar system. Gravity pulled the gas and dust together to form the planet. As soon as it formed, Earth traveled around the Sun in an almost-circular orbit.

The early solar system was messy. It had lots of large rocks and debris flying around in it. Some of the rocks, called planetesimals, were huge. Scientists now think that one of these planetesimals, the size of Mars, started heading for Earth.

Imagine you had been on Earth to witness the event. The planetismal first appeared as a dot in the sky. Over a period of days and weeks, it appeared bigger and bigger. Then it completely blocked the view from Earth in that direction. Finally, it struck, traveling at perhaps 40,000 kilometers (km) per hour. The collision lasted several minutes.

The crash created a chain of events. First, the impact seemed to destroy the incoming object. The planetesimal turned to gas, dust, and a few large chunks of rock. Some surviving chunks traveled deep into the interior of Earth. A large portion of Earth was destroyed as well. The energy that resulted from the crash produced a huge explosion. Earth itself might have been in danger of being blasted apart.

Second, the explosion threw a tremendous amount of matter into motion. Some of this debris flew far out into space. Other matter flew up into the air and then returned to Earth. This debris came in many sizes. Some of it was huge rocks that immediately returned to Earth. A short time later, smaller granules of different sizes fell to Earth. Months or even years later, some of the debris was still floating up in the air.

A large portion of the debris didn't fly off into space, and it didn't return to Earth. It began orbiting Earth. This orbiting debris formed a disk, like the rings of Saturn. The ring was probably about two Earth diameters from the surface of Earth. Over several years, the pieces of debris started to attract one another. Gradually, they formed into larger and larger chunks. Eventually, the chunks of debris formed Earth's Moon.

Earth now had a satellite where previously there was none. It must have been quite a sight up there only about 30,000 km above Earth. Today the Moon is much farther above Earth, about 385,000 km.

A representation of how the Moon might have formed

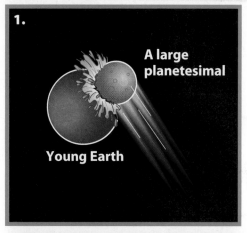

1. A large planetesimal

Young Earth

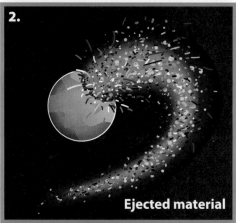

2. Ejected material

3. Orbiting debris formed a disk.

4. Earth

Debris formed the Moon.

Stargazing

Stars are twinkling points of light in the night sky. When you get into bed at night, the sky is filled with stars. But in the morning, they are gone. Where did they go?

The stars didn't go anywhere. They are exactly where they were when you went to sleep. But you can't see the stars in the day sky. This is because the light from our star, the Sun, is so bright.

Where Are the Stars?

Stars are huge balls of hot gas. Most stars are located in groups of stars called **galaxies**. The Sun is in the galaxy called the Milky Way. There are several hundred billion other stars in the Milky Way with us. If we could see the entire Milky Way from above, it might look something like the picture below. The Sun is out on one of the arms where the arrow is pointing.

The Sun is one of the billions of stars in the Milky Way.

As you can see, we are surrounded by stars. Think about the 2,000 or so stars you can see and the billions of stars you can't see in the Milky Way. All these stars, including the Sun, are moving slowly around in a huge circle. Because all the stars move together, the positions of the stars never change. You can see the same stars in the same places in the sky year after year.

Did you ever see the **Big Dipper**? It is seven bright stars in the shape of a dipper. The Big Dipper is part of a **constellation** called Ursa Major, or the Great Bear.

Most of the stars you see in the night sky are part of a constellation. A constellation is a group of stars in a pattern. Thousands of years ago, stargazers imagined they could see animals and people in the star groups. They gave names to these constellations. Some of the names are Orion the hunter, Scorpius the scorpion, Aquila the eagle, Leo the lion, and Gemini the twins. Those same constellations are still seen in the sky today. They are unchanged.

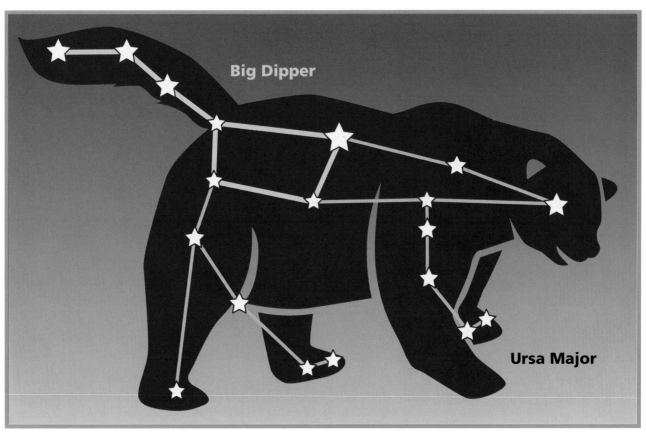

The constellation Ursa Major (the Great Bear)

Constellations in Motion

Even though the stars don't change position, they appear to move across the night sky. Stars move across the sky for the same reason that the Sun and the Moon move across the sky. The stars are not moving. Earth is moving. As Earth rotates on its **axis**, constellations rise in the east. They travel across the night sky and set in the west.

If you look at the stars every day for 1 year, you will see something interesting. The stars you see in winter are different from the stars you see in summer. If the stars don't move around, how is that possible? To answer this, we have to look at how Earth orbits the Sun.

Here is a simple drawing of the Milky Way. The Sun and Earth appear much larger than they really are. That's so we can study what happens as the seasons go by.

The side of Earth facing the Sun is always in daylight. The side facing away from the Sun is always in darkness. You can only see stars when you are on the dark half of Earth.

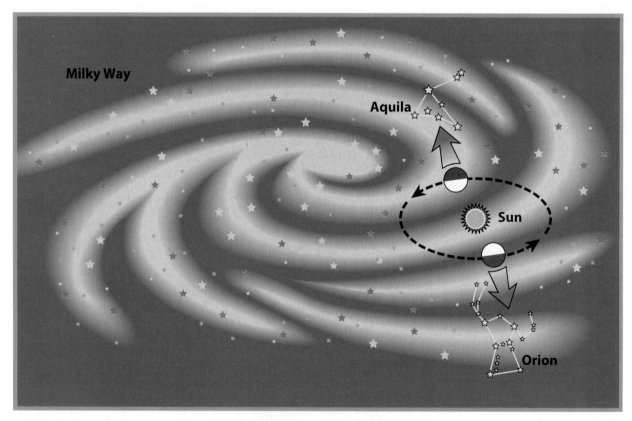

A simple drawing of the Sun ☼ **and Earth** ◗**, not drawn to scale.**

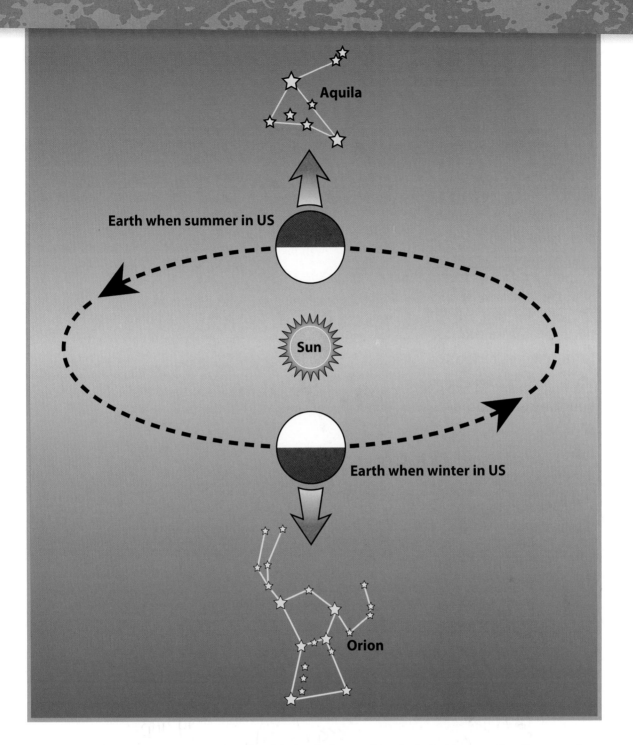

When it is summer in the United States, Earth is between the Sun and the center of the Milky Way. The constellation Aquila is in that direction. The dark side of Earth is toward the center of the galaxy in summer. On a summer night, you see Aquila high overhead.

Six months later, Earth is on the other side of the Sun. It is winter in the United States. Now the dark side of Earth faces away from the center of the galaxy. The constellation Orion is in that direction. On a winter night, you see Orion high overhead.

This is Orion. Can you see his belt and sword? The brightest stars in the Orion constellation appear in this picture. You can see Orion in the sky on a clear winter night.

When you see Orion, you are seeing the same pattern of stars that a hundred generations of stargazers looked at before you. And a hundred generations into the future, stargazers will still see Orion marching across the winter sky.

The constellation Orion is visible in the winter sky.

Thinking about the Stars

1. Why do stars appear to move across the night sky?

2. What is a constellation?

3. Why are the constellations in the summer sky different from those in the winter sky?

4. Imagine that you could see stars during the day. What constellation would you see at noon in winter? Why do you think so?

Looking through Telescopes

What do you see when you look at the sky on a clear night? You probably see many twinkling stars. Maybe you see the Moon or a planet. People saw the same objects in the sky thousands of years ago.

The way we look at objects in the sky changed in 1608. In that year, the **telescope** was invented. A telescope is a tool that **magnifies** distant objects so that they appear larger and closer.

Galileo Galilei (1564–1642) was a scientist who lived in Italy. In 1609, he improved the telescope and used it to observe the night sky. He could see many more stars through

Galileo Galilei

the telescope than with his unaided eyes. He could see mountains and craters on the Moon. And he could see that planets were spheres, not just points of light. Then Galileo turned his telescope toward Jupiter. He became the first person to observe moons orbiting another planet.

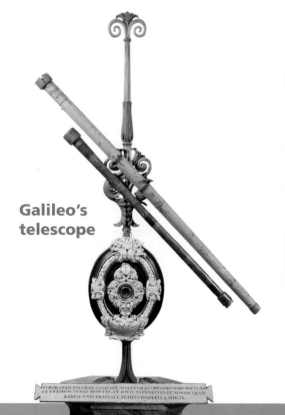

Galileo's telescope

You can see more stars through a telescope than with your unaided eyes.

63

The Apollo 11 landing site on the Moon

As telescopes got more powerful, astronomers could see more details on planets. They could also see more stars in the night sky. By the mid-1900s, the surface of the Moon could be studied in detail with telescopes on Earth. Scientists used pictures taken through telescopes to plan the first Moon landing in 1969.

Modern Telescopes

Most telescopes are built on mountain peaks. The telescopes are above most of the dust and pollution in the air. And they are far away from city lights. The telescopes are protected inside dome-shaped buildings called **observatories**.

Keck Observatory is on top of Mauna Kea, a 4,205-meter peak on the island of Hawaii.

The space shuttle placed a very important telescope in Earth's orbit in 1990. It is called the Hubble Space Telescope. The Hubble Space Telescope takes pictures of planets and other objects in the solar system. It also takes pictures of objects beyond the solar system. Because the telescope orbits above Earth's atmosphere, it gets a clear view of outer space.

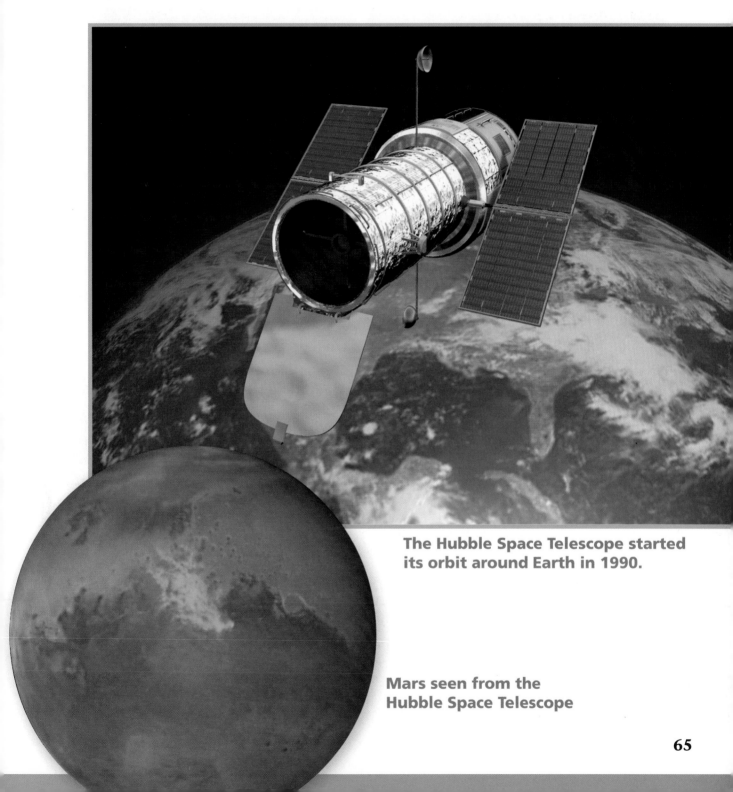

The Hubble Space Telescope started its orbit around Earth in 1990.

Mars seen from the Hubble Space Telescope

When you look up at the sky on a clear, moonless night, you can see about 2,000 stars. But that view changes a lot when you look through the Hubble Space Telescope. You can see millions of stars that are too dim to see with your unaided eyes. Telescopes make distant objects look bigger and closer. With telescopes, astronomers can explore space without leaving Earth.

Part of the Milky Way seen with unaided eyes on a clear night

Part of the Milky Way seen through the Hubble Space Telescope

Thinking about Telescopes

1. Who was Galileo, and what was he the first to do?

2. Why is a telescope a useful tool to an astronomer?

3. Why are modern telescopes built on mountaintops or put into space?

Star Scientists

Sometimes a childhood fascination with stars lasts a lifetime. Scientists who try to find out the secrets of stars are called astronomers. Meet five scientists who have taken star study in different directions. They truly are star scientists.

Stephen Hawking

When you drop a ball on Earth, the force of gravity pulls it down. Gravity keeps your feet on the ground, too. When a star reaches the end of its life, it collapses. Gravity pulls together all the matter in the star. In collapsed stars, gravity can even pull in light.

When a really big star collapses, it can become a **black hole**. In a black hole, gravity is so strong that nothing, not even light, can escape. Everything for millions of kilometers (km) around is pulled into the black hole, where it disappears.

Today, the best-known scientist who studies black holes is Stephen Hawking (1942–). Using mathematics, Hawking helped prove that black holes exist.

Since 1994, the Hubble Space Telescope has been used to search

Stephen Hawking

for evidence of black holes. Hubble images show stars and gases swirling toward a central point. Hawking says this could be the effect of a black hole. A black hole's strong gravity would pull in everything around it, including stars. Future images from the Hubble Space Telescope and its successor, the James Webb Space Telescope, might help scientists improve their understanding of black holes.

Edna DeVore

Edna DeVore

Many people wonder if there is life anyplace else in the universe. But Edna DeVore (1947–) does more than wonder. DeVore is the Deputy Chief Executive Officer (CEO) of the SETI Institute. SETI is short for Search for Extraterrestrial Intelligence.

The scientists at SETI think that there might be other intelligent beings in the universe. If they are out there, they live on a planet orbiting a star. And intelligent life could develop technologies that send signals into space. Radio, TV, navigation systems, and telephones on Earth send messages in all directions into space. Is someone else out there doing the same thing?

The SETI Institute watches the sky for any signs of life in the universe. It uses big sets of antennae to listen for any sounds of life, like radio signals.

DeVore is a scientist and educator at SETI. She grew up on a ranch in Sattley, California. DeVore remembers watching the stars and the Milky Way in the clear night sky as a child. She didn't think about becoming a star scientist, but in college, DeVore became more and more interested in the stars. After getting her degree in **astronomy**, she became a teacher and a **planetarium** director. But the question she always asked herself was "Are we alone in the universe?"

DeVore is in charge of education and public outreach for the SETI Institute and NASA's Kepler Mission. And what's the latest report from the universe? The scientists at SETI haven't heard or seen anything yet. But they keep watching and listening.

SETI uses radio telescopes like this one.

Neil deGrasse Tyson

You put water and fish in an aquarium. You put soil and plants in a terrarium. But what do you put in a planetarium? Planets! A planetarium is filled with information about planets, stars, galaxies, and everything else seen in the night sky.

Neil deGrasse Tyson

A planetarium is a theater with a dome-shaped ceiling. In the middle of the room is a projector. The projector shines points of light all over the dome. The points of light are in the same positions as the stars in the sky. The projected stars make it seem as though you are outside watching the stars.

One of the fun things about a planetarium is that you can control the night sky. Do you want to see the stars as they were the day you were born? Or how the sky looked at different times in Earth's history? The projector operator can put you under the stars at any time and any place.

When he was a child, Neil deGrasse Tyson (1958–) never dreamed that he would one day be in charge of a planetarium. Tyson took a class at the Hayden Planetarium in New York City when he was in middle school. He was awarded a certificate at the end of the class. It meant a lot to him.

Tyson's love of the stars grew as he got older. After getting a PhD in astrophysics, Tyson spent time doing research and promoting education. He researched how stars form and explained space science to the public. He works hard to make science interesting for everyone. In 1996, Tyson became the youngest person ever to direct the Hayden Planetarium. It is the same place he visited as a child.

The Hayden Planetarium

Mae Jemison

Mae Jemison (1956–) was born in Decatur, Alabama. She moved to Chicago, Illinois, as a child. There an uncle introduced her to astronomy. In high school, Jemison began reading books on astronomy and space travel. She was only 16 years old when she entered college. She earned degrees in chemical engineering and African and Afro-American studies from Stanford University. She went on to earn her medical degree from Cornell University.

Mae Jemison, astronaut

After becoming a doctor, Jemison spent time in western Africa as a Peace Corps physician. But she continued to think about astronomy and space travel. She wanted to be part of the space program.

Jemison joined the astronaut program in 1987. On September 12, 1992, Jemison became the first African American woman in space. She was a science mission specialist on the space shuttle *Endeavour.*

Jemison conducted experiments to find out more about the effects of being in space. She studied motion sickness, calcium loss in bones, and weightlessness.

The crew of *Endeavour*

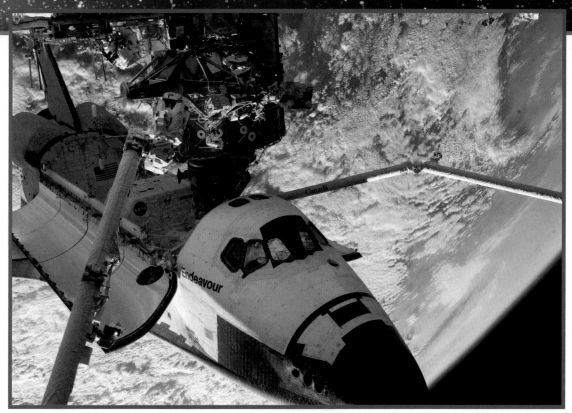

The space shuttle *Endeavour* docked at the International Space Station

Space shuttle mission STS-47 was the 50th space shuttle flight, but only the second flight for *Endeavour*. The space shuttle was in space for 8 days. During those 8 days, Jemison orbited Earth 127 times at an altitude of 307 km. The space shuttle traveled 5,234,950 km.

In 2011, after 30 years of flying and many firsts, the space shuttle program ended. Did the space shuttles actually fly in space? No, they orbited Earth in the upper atmosphere. What kept the shuttles in orbit? The answer is gravity. Shuttles traveled very fast. Earth's gravity pulled on the shuttles, constantly changing their direction of travel. Engineers from NASA figured out exactly how high above Earth's surface and how fast the shuttles needed to travel. Since they knew the force of gravity, the space shuttles were able to stay in orbit until the astronauts changed the speed. Then gravity pulled them back to Earth.

Ramon E. Lopez

As strange as it may sound, there is weather in space. But it's not weather like we have on Earth. There are no clouds, hurricanes, or snowstorms in space. Space weather is the result of activities on the Sun. The Sun is always radiating energy into the solar system. The regular flow of light and gases is called **solar wind**. But what happens when the Sun goes through a period of violent solar flares? That's what Ramon E. Lopez (1959–) studies.

Ramon E. Lopez

Solar flares are huge solar explosions. They send intense blasts of electrified gas into Earth's atmosphere. The blasts can produce electric effects in the atmosphere and on Earth's surface. The electricity can disable satellites orbiting Earth and interfere with radio transmissions and cell phones. Space weather can cause blackouts over large areas.

Lopez and his team understand how space weather can damage communication and navigation systems. And they understand how important these systems are to modern society. Can Lopez and his team learn how to predict space weather? Will it be possible to warn the world when a dangerous solar storm is coming? Lopez and the team he works with are developing a computer program to predict space weather about 30 minutes before it hits Earth. And that may be just long enough to protect communication and navigation systems from damage.

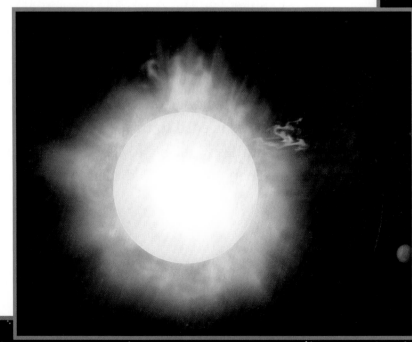

The Sun with a large solar flare

Our Galaxy

Stars are huge balls of hot gas. They produce bright light that travels out into space. When we go outdoors on a clear night, we see the stars as tiny points of light. There are billions of them sending light our way. But because most of them are so far away, the light is too dim for us to see. We can enjoy the 2,000 or so stars that we can see with our unaided eyes.

Astronomers study stars and other objects in the sky. One of the most important tools they use is the telescope. Telescopes magnify objects in the sky. When an astronomer looks at an object through a telescope, the object looks bigger and closer. With a telescope, you can see many more stars. Objects in the night sky can be studied in greater detail with a telescope.

The great Italian scientist Galileo Galilei used a telescope to observe the Moon and planets. He saw things no one had seen before. He observed mountains and craters on the Moon and discovered moons orbiting the planet Jupiter. Galileo's telescope brought the science of astronomy to a new level.

Galileo's record of the movement of Jupiter's moons

Galileo's painting of the phases of the Moon

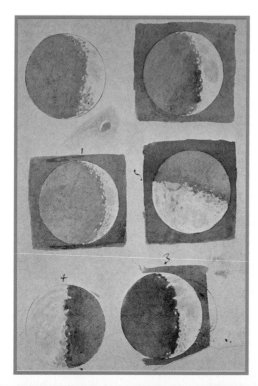

73

Moving Objects in the Sky

The Sun, the Moon, and the stars all move in the sky. But they all move in a different way. The Sun rises in the east in the morning and sets in the west at night. We see the Sun only during the day, never at night. Every day the Sun has the same pattern.

The Moon rises in the east and sets in the west, just like the Sun. But it doesn't always rise and set at the same time. Sometimes it rises in the morning, and sometimes it rises in the afternoon. There is a chance you might see the Moon in the daytime, but it might appear at night, depending on the lunar cycle.

The Moon and the Sun rise in the east and set in the west because Earth rotates on its axis. To people on Earth, the Sun and the Moon appear to move across the sky. The Moon rises and sets at different times because the Moon is orbiting Earth. The Moon is changing its position all the time.

Stars are different. We see them only at night. They are up in the sky all the time. But we can't see them during the day because the Sun is too bright. As soon as the Sun sets, the sky gets dark. Then we can see the stars. Stars rise and set, too. If you watch one star, you can see it rise above the eastern horizon. It then appears to move across the night sky, and set in the west. Why? Because Earth is rotating.

Earth's Orbit

One more thing is different about stars. Earth is completely surrounded by stars in all directions. But you can't see all of them at once. This is because half of them are on the day side of Earth. Also, the stars you can see in winter are different from the ones you can see in summer.

Here's why. Earth orbits the Sun. One complete orbit takes a year. At all times, half of Earth is in daylight and half is in darkness. It is always the side of Earth toward the Sun that is in daylight. The day side of Earth is always "looking" toward the Sun.

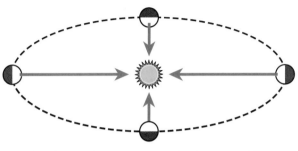

Half of Earth is in daylight at all times.

The side of Earth away from the Sun is always in darkness. The dark side of Earth is always "looking" away from the Sun. We can only see stars at night when it is dark. So stargazers always look in the direction away from the Sun.

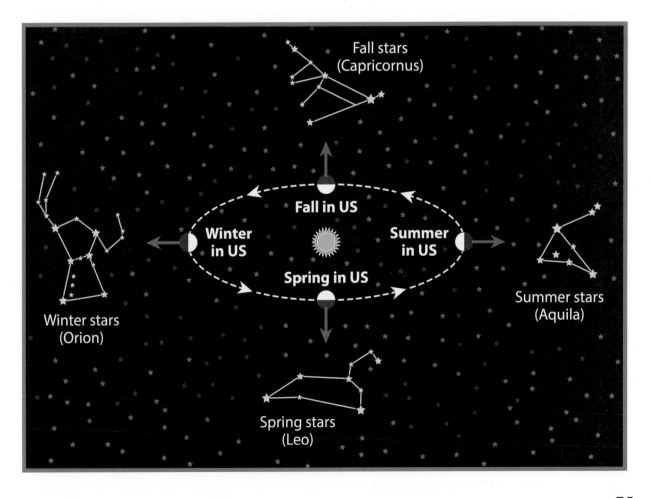

Fall stars
(Capricornus)

Fall in US

Winter in US

Summer in US

Spring in US

Winter stars
(Orion)

Summer stars
(Aquila)

Spring stars
(Leo)

As Earth orbits the Sun, the dark side of Earth is aimed at different parts of the star-filled sky. Because stars don't move, the stars and constellations you see change from season to season.

In the United States, look for Aquila the eagle in summer. Look for Capricornus the goat in the fall, Orion the hunter in winter, and Leo the lion in spring.

Leo the lion

Science Safety Rules

1. Listen carefully to your teacher's instructions. Follow all directions. Ask questions if you don't know what to do.

2. Tell your teacher if you have any allergies.

3. Never put any materials in your mouth. Do not taste anything unless your teacher tells you to do so.

4. Never smell any unknown material. If your teacher tells you to smell something, wave your hand over the material to bring the smell toward your nose.

5. Do not touch your face, mouth, ears, eyes, or nose while working with chemicals, plants, or animals.

6. Always protect your eyes. Wear safety goggles when necessary. Tell your teacher if you wear contact lenses.

7. Always wash your hands with soap and warm water after handling chemicals, plants, or animals.

8. Never mix any chemicals unless your teacher tells you to do so.

9. Report all spills, accidents, and injuries to your teacher.

10. Treat animals with respect, caution, and consideration.

11. Clean up your work space after each investigation.

12. Act responsibly during all science activities.

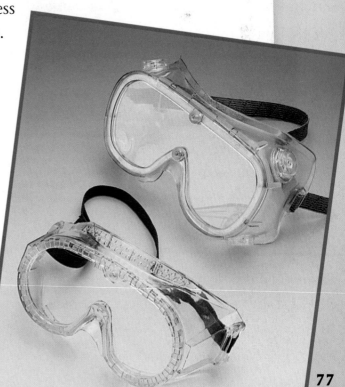

Glossary

asteroid a small, solid object that orbits the Sun

astronomer a scientist who studies objects in the universe, including stars, planets, and moons

astronomy the study of the universe and the objects in it

atmosphere the layer of gases surrounding Earth

axis an imaginary line around which a sphere, like a planet, rotates

Big Dipper a group of seven bright stars in the shape of a dipper

black hole a region in space without light that has a strong gravitational pull

comet a mass of ice, rock, and gas orbiting the Sun

constellation a group of stars that humans see as a pattern and give a name

crater a hole formed by an object hitting a surface

crescent Moon the curved shape of the visible part of the Moon just before and after a new Moon

cycle a set of events or actions that repeat in a predictable pattern

day the time between sunrise and sunset on Earth

diameter the straight-line distance through the center of an object, one side to the other side

dwarf planet a round object that orbits the Sun but does not orbit a planet

Earth the third planet from the Sun

extraterrestrial beyond Earth

first-quarter Moon a phase of the Moon in the lunar cycle halfway between a new Moon and a full Moon

full Moon the phase of the Moon when all of the sunlit side of the Moon is visible from Earth

galaxy a group of billions of stars. Earth is in the Milky Way galaxy.

gas giant planet one of the four planets that are made of gas. These are Jupiter, Saturn, Uranus, and Neptune.

gibbous Moon the shape of the Moon when it appears to be more than a quarter but not yet full and when it is less than full but not quite a third quarter.

gravitational attraction the mutual force pulling together all objects that have mass

gravity the force of attraction between two objects

Kuiper Belt a huge region beyond the gas giant planets, made up of different-size icy chunks of matter

lunar cycle the 4-week period during which the Moon orbits Earth one time and goes through all of its phases

lunar eclipse when Earth passes exactly between the Moon and the Sun

magnify to make an object appear larger

mass the amount of material in something

matter anything that takes up space and has mass

Milky Way the galaxy in which the solar system resides

Moon Earth's natural satellite

new Moon the phase of the Moon when the sunlit side of the Moon is not visible from Earth

night the time between sunset and sunrise on Earth

observatory a building that protects telescopes

orbit to move or travel around an object in a curved path

phase the shape of the visible part of the Moon

planet a large, round object orbiting a star

planetarium a theater with a dome-shaped ceiling that represents the sky

predict to estimate accurately in advance based on a pattern or previous knowledge

reflect to bounce off an object or surface

rotate to turn on an axis

satellite an object, such as a moon, that orbits another object, such as a planet

season a time of year that brings predictable weather conditions to a region on Earth

shadow the dark area behind an object that blocks light

solar eclipse when the Moon passes exactly between Earth and the Sun

solar system the Sun and the planets and other objects that orbit the Sun

solar wind the regular flow of particles from the Sun

star a huge sphere of hydrogen and helium gas that radiates heat and light

Sun the star at the center of the solar system around which everything else orbits

telescope an optical instrument that makes objects appear closer and larger

terrestrial planet one of the four small, rocky planets closest to the Sun. These are Mercury, Venus, Earth, and Mars.

thermonuclear reaction a change in atomic structure that creates heat and light energy, such as the reactions that occur on the Sun

third-quarter Moon a phase of the Moon in the lunar cycle halfway between the full Moon and the new Moon

unaided eyes looking at something without the use of a telescope or microscope

waning getting smaller

waxing getting larger

Index